BASE TEN MATHEMATICS

Interludes for Every Math Text

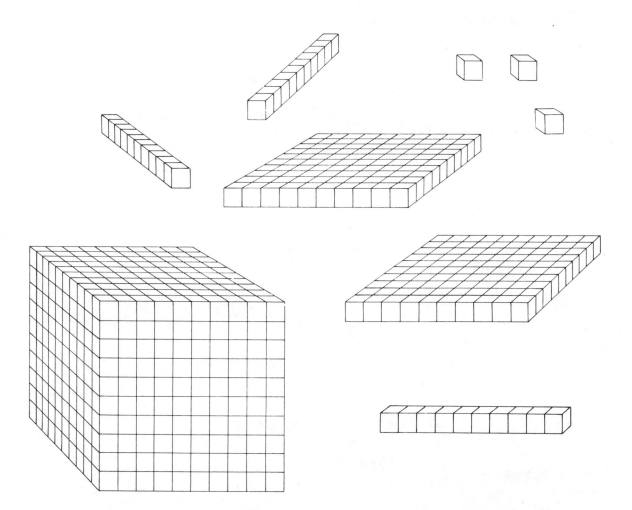

by Mary Laycock

© ACTIVITY RESOURCES COMPANY, INC., Box 4875, Hayward, CA 94545

INTRODUCTION

These activities are designed as concept developing INTERLUDES to be used with children as they make use of the BASE TEN BLOCKS. The activities demonstrate how the blocks are a TOOL to be used for solving problems in the child's regular mathematics program from grade one through junior high. Rather than grade specific, the activities are keyed to a PREREQUISITE experience. Most can be pursued and practiced using the problems from any good math text so the ones given are simply starters or examples of the strategies to be used to help students solve their own problems using base ten blocks. A good test of a teacher's effective use of the blocks is that students seek the blocks to solve their problems BEFORE they seek the teacher's help. If they do not the teacher should respond with, "Let's see how it works; come let's build it."

Materials needed for each activity are listed. Illustrations have been used as often as possible. Some of the kinds of questions that produce the most thinking and understanding have been included. It is always intended that blocks be used to work out problems. Abbreviated forms like large squares for flats in presenting rectangular arrays are for the teacher's information; children should work out the responses with the concrete material and record with numerals.

As often as feasible the pages have been arranged for use by the child, but the necessity of communicating with the teacher has demanded shorter, more concise forms on some pages.

Mary Laycock
Mathematics Specialist
Nueva Learning Center
Hillsborough, California 94010

What do we want children to learn in Mathematics? Basic skills, certainly—skills in counting; reading and writing numerals; computing sums, differences, products, and quotients; solving equations, etc.—but not these skills exclusively. The "drills on skills" approach to Mathematics is narrow; it omits, or underemphasizes the development of powers of mathematical *reasoning*, and reasoning is the very essence of mathematics.

Mechanical computation skills are not enough for solving problems in the real world. When mathematics is involved, one needs to know which relationship, pattern, operation, or formula is appropriate, and whether the answer to the problem makes sense.

Understanding and *reasoning* go together. The work of Piaget and other psychologists indicate that the child develops understanding and powers of reasoning through exploration, interaction, and feedback. Manipulative aids can serve as a medium for this process. One of the most useful aids in the elementary school Math program is a set of Base Ten Blocks. These blocks offer a clear and logical model for the place value system of numeration and for each step of an arithmetic computation process, as well as for basic algebra and metric system concepts.

Mary Laycock is an expert in the use of manipulative aids in teaching mathematics. Base Ten Blocks are among her favorites. She has shown in her work with children at Nueva Day School and in her professional development sessions with teachers that basic Math skills can be learned with greater facility and understanding through the judicious use of these blocks. In this book she shares some of her secrets. Here you will find more than 50 basic activities that span a wide range of major topics in the Grades 1-8 Math program.

The activities in this book are based on a pattern of "Build—Process—Record" that will help students to work independently and with understanding. A feature of the Base Ten Blocks Activities Program is that it is designed for co-ordination *with any school, district, or state adopted textbooks.* An added feature of the blocks is their metric attributes. A "unit" is a model of a cubic centimeter and it has a mass of one gram. A "long" is 10 centimeters, or 1 decimeter in length. A "flat" can serve as a model for both an area of 100 square centimeters and a volume of 100 cubic centimeters, and a stack of 10 "flats" or 1 "block" has a volume of 1000 cubic centimeters or 1 liter.

The Base Ten Blocks Program won't do the whole job; a variety of materials and applications is desirable in any Math program. But the principles that can be developed with the help of the blocks are basic, and the suggestions given here can be modified for use with other materials, such as beansticks, Cuisenaire Rods, abacuses, etc. Whether you teach Old Math, New Math, or In-between Math, Base Ten Blocks make a perfect fit for your program.

Joseph Moray
Professor of Education
San Francisco State University

TABLE OF CONTENTS

© ACTIVITY RESOURCES COMPANY, INC., Box 4875, Hayward, CA 94545

BUILDING AND INVESTIGATING

PREREQUISITE: None
MATERIALS: Base Ten Blocks
PROCEDURE: Make availible a collection of base ten blocks and encourage students of any age to build. Sometimes an ivitation like, "Who can build the highest structure with the fewest blocks?" or "Build a castle and tell me how much it is worth in units (for older students)."

MATCH AND FIND OUT

PREREQUISITE: Time to have played.

MATERIALS: Base Ten Blocks

PROCEDURE: Do not answer until you have matched to find out.

How many units make a long?

unit   long   _____ units = 1 long

How many longs make a flat?

_____ longs = 1 flat

How many flats make a block?

flat

block

_____ flats = 1 block

Use base ten blocks to answer these questions:

_____ units = 3 longs   _____ longs = 5 flats

_____ longs = 2 flats   _____ longs = 1 block

_____ units = 1 flat   _____ units = 4 longs

_____ flats = 4 blocks   _____ units = 2 flats

2

COUNTING STRIP—THE MOST ESSENTIAL ACTIVITY

PREREQUISITE: Opportunity to play with the blocks, numeral recognition of 0, 1, 2, 3, 4, 5, 6, 7, 8, and 9. Introductory experience with numeral formation; the skill will come as a result of the activity.

WARNING: Do not try to start a whole class at the same time—only as many children as can have close supervision in the trading process and numeral formation.

MATERIALS: Base ten blocks, adding machine tape, and a piece of construction paper twice the size of this page marked as the diagram below.

PROCEDURE: The child places a unit on the mat in the units place and records 1, another unit and records 2, another unit and records 3, etc. When ten are there, the pencil goes down until the child matches the ten units to the long to make sure the trade for a long is fair. Ten is verbalized as one long and no units.

MY COMPUTER			
blocks	flats	longs	units

As each decade is completed the child matches to make sure the trade is fair. This should continue until the child is in the thirties or forties or until the alloted time for the lesson expires. Next day, the child reads the tape, for example 47, and gets out four longs and seven units and continues, each time getting out another unit. At 99 help should be available to make the two trades needed and record, 1 flat, 0 longs, 0 units. Do this to at least 100. The day-after-day building of a strip to 1000 makes place value and numeral formation automatic and understandable. Children often beg to go past 3000. Careful supervision to assure correct numeral formation pays rich dividends in efficiency later. (Never again will a child work so hard for mastery. Be sure to celebrate and show enthusiasm for their progress!)

1
2
3
4
5
6
7
8
9
10
11
12
13
14
.
.
.
.

SPIN A FLAT

Game for four—three players and a banker

PREREQUISITE: Experience with the counting strip; sums to ten, not mastered.

MATERIAL: Base ten blocks, die, and computer mats for three children.

PROCEDURE: One child is banker. Each child spins the die to find the highest, who goes first. Play continues clockwise. On each spin, the child asks the banker for the number of units indicated and places them on the units space of the computer. When there are ten, the child matches and asks the banker for a long. Whoever trades ten longs for a flat first wins.

MY COMPUTER			
blocks	flats	longs	units

GO BROKE

PREREQUISITE: Experience of playing Spin a Flat.

MATERIALS: Base ten blocks, die, three computer mats.

PROCEDURE: Each player begins with a flat. One child is banker; the other three spin the die to see who spins the lowest. The lowest goes first; play proceeds clockwise. Each child spins, the number spun must be removed. The banker will have to be asked for change so the removal can be accomplished. The winner is the first to reach zero exactly. (The number of the last spin must be the exact number of blocks remaining.)

4

HUNGRY BUG ADDITION

PREREQUISITE: Experience with number combinations to ten (mastery not needed) and progress to 100 on the counting strip.

MATERIAL: Base Ten Blocks.

PROCEDURE: (To develop strategies for the sums between ten and twenty the "Hungry Bug" who wants to be ten help them reason before it is necessary to remember.)

To solve **9**
         + 4   Take out nine units and line them up beside a long:

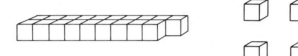

Nine is a "Hungry Bug" that wants to be ten; how many more does he need? (One) If he gobbles up one of the four, what is left? (3) The sum is 10 + 3.

What happens to the Hungry Nine Bug? Use a long to measure and see:

9	9	9	9	9	9	9	9
+2	+5	+7	+3	+9	+6	+4	+8

What is the rule?

Eight is a Hungry Bug with an appetite for how much?   (yes, two.)

To solve: **8**
          + 3   Take out eight units and line them up beside a long:

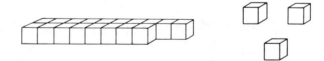

Eight is a "Hungry Bug" that wants to be ten; how many more does he need? (Two)   If he gobbles up two of the three, how many are left?

8	8	8	8	8	8	8	8
+3	+5	+7	+9	+2	+4	+6	+8

Seven is a Hungry Number Bug that is even hungrier. Can you find out about 7?

COMPARISON

PREREQUISITE: Experience with the base ten counting strip past 100.

MATERIAL: Base Ten Blocks

PROCEDURE: Build each of the numbers from longs and units, write >, <, or =.

A.

_____ > _____

B.

_____   _____

C.

_____   _____

Find more problems like this in your book; build and decide. (Teachers: Students may need help in identifying the appropriate pages in their texts.)

6

COMPARISON ACTIVITY FOR A GROUP OF TEN

PREREQUISITE: Experience of writing the base ten strip to about 300.

MATERIAL: Base Ten Blocks

DIRECTIONS: Duplicate pages 8 through 11. Cut apart each picture of blocks. Give each child in the group one picture to build out of blocks. Also, give each child a copy of the chart below to record the numbers describing all the block piles of the children in the group. After all have finished ask the children to arrange their piles of blocks in a line from smallest to largest. (Follow-up activities where each child is handed a large number and asked to build it. They can be compared in the same way.)

A. _____   F. _____

B. _____   G. _____

C. _____   H. _____

D. _____   I. _____

E. _____   J. _____

Alternate activity: Give each child a card with a different number on it and ask them to build it. Once built their card should be hidden so the other class members have to examine the blocks and record the number on a copy of the form above.

A. 3450	D. 5430	G. 354	I. 4503
B. 3045	E. 5034	H. 4035	J. 4053
C. 3405	F. 5403		

A.

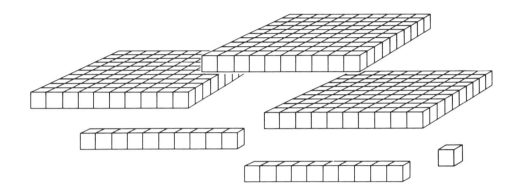

B.

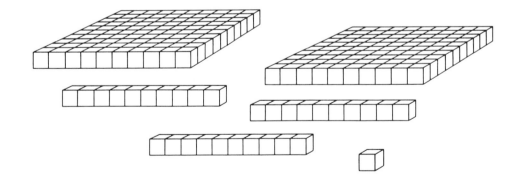

C.

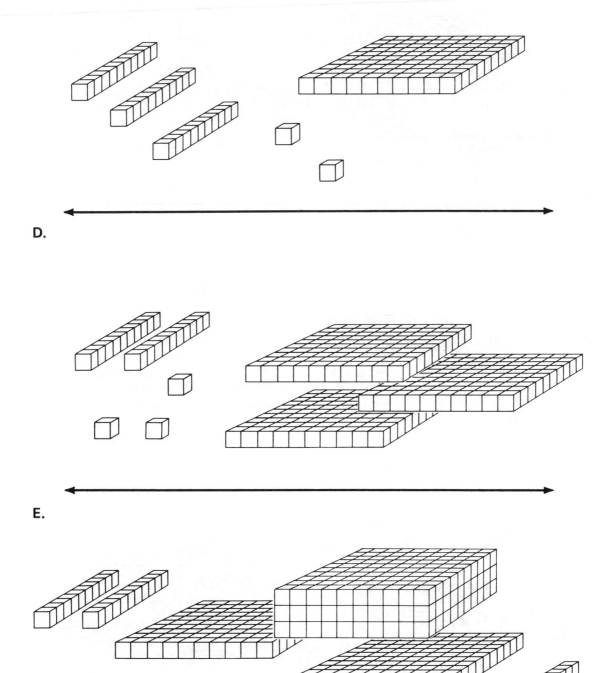

D.

E.

F.

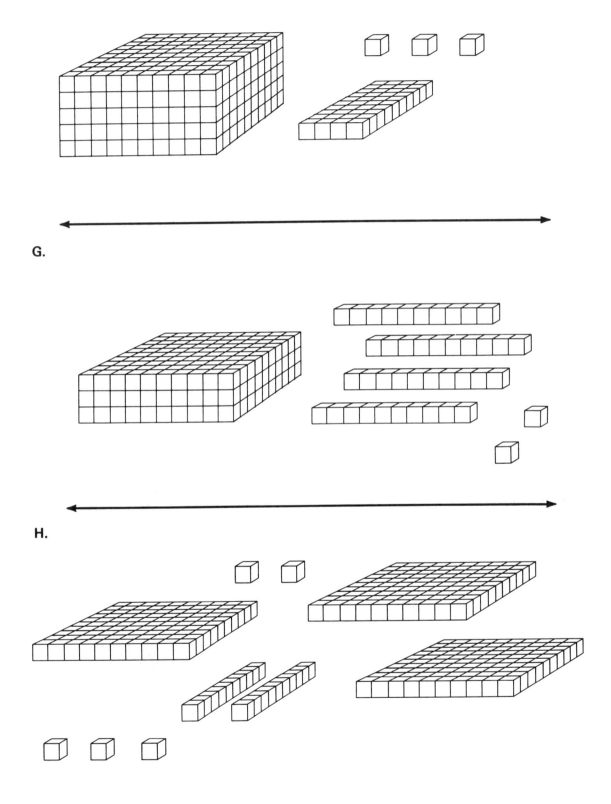

G.

H.

I.

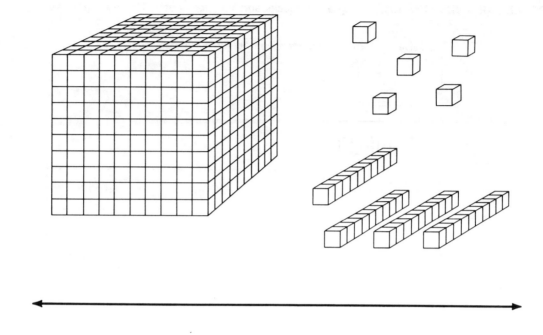

J.

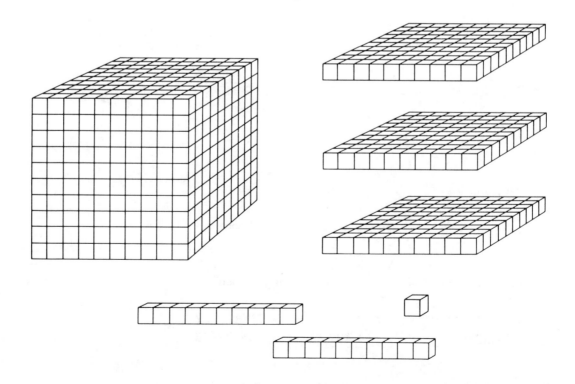

PREREQUISITE: Experience with adding two one-digit numbers or Hungry Bug Addition.

PROCEDURE AND MATERIALS: Base ten blocks and your computer. No regrouping yet.

To solve $\begin{array}{r} 24 \\ + 35 \end{array}$ build each number on your computer.

MY COMPUTER

	longs	units

How many units?   Can you Trade?   Record.

How many longs?   Can you trade?   Record.

Do these by building. There are more like them in your book.

1.	$\begin{array}{r} 43 \\ +55 \end{array}$	2.	$\begin{array}{r} 55 \\ +43 \end{array}$	3.	$\begin{array}{r} 82 \\ +16 \end{array}$	4.	$\begin{array}{r} 16 \\ +82 \end{array}$

Which problems are alike?   How are they different?

12

ADDITION

PREREQUISITE: Experience on the number strip to 100. Played Spin a Flat Game.

MATERIAL: Base Ten Blocks.

PROCEDURE: Build these numbers. Hungry Bug addition means to push the two piles together, trade 10 units for a long, and record your result.

Build   **346**
        + 26

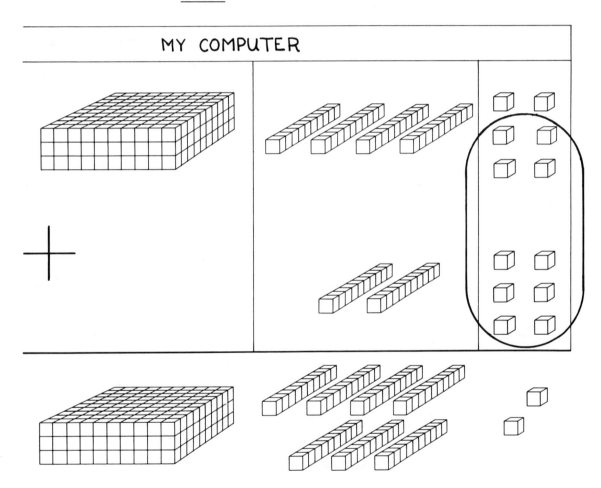

MY COMPUTER

Build and answer these and those like them from your book:

1.	457	2.	456	3.	58	4.	159	5.	260	6.	361
	+ 35		+ 35		+ 35		+ 35		+ 35		+ 35

13

PREREQUISITE: Putting it together addition. Hungry Bug Addition.

PROCEDURE AND MATERIALS: Base ten blocks and your computer. Build the problem.

PROBLEM: 65
           + 58

MY COMPUTER

1.   How many units? (13)   Can you trade? (Yes)
2.   How many units left? (3)   Record.
3.   Carry the long to the tens place. Record.
4.   How many longs all together? (12)   Can you trade? (Yes)   Record.

Find more problems like this in your book. Add with the units first and make a trade when possible, then add the longs and trade when possible.

1. 74   2. 74   3. 74   4. 74
   + 18   + 19   + 29   + 28

14

ADDING BIG NUMBERS

PREREQUISITE: Addition (Watch out for a Trade)

MATERIALS: Base ten blocks, computer mats

PROBLEM: Build   263
                         +478

MY COMPUTER

How many units?   Trade?   Record.

How many longs?   Trade?   Record.

How many flats?   Trade?   Record

$$\begin{array}{r} {}^{1}\ {}^{1}\ \\ 2\ 6\ 3 \\ +4\ 7\ 8 \\ \hline 7\ 4\ 1 \end{array}$$

Find more like these in your book. Build and record.

1. 263	2. 263	3. 263	4. 263
+ 778	+ 777	+ 737	+ 728

15

PREREQUISITE: Experience on the base ten strip to 100. Played Go Broke. Hungry Bug Addition.

MATERIALS: Base ten blocks, computer mats.

PROCEDURE: (The purpose here is to help students use strategies for subtraction rather than memorized facts.)

PROBLEM:
$$\begin{array}{r} 14 \\ -\ 6 \\ \hline \end{array}$$

STEP 1.   Build 14:

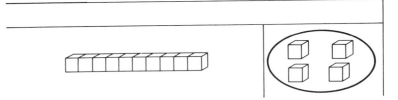

STEP 2.   Take away 4 to leave 10:

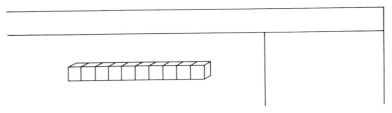

STEP 3.   Trade the long for 10 units; how many units must be removed so the 6 has been taken away? (2 more)   Record the number left.

$$\begin{array}{r} 14 \\ -\ 6 \\ \hline 8 \end{array}$$

Find more problems like these in your book. Build with blocks and follow these steps:

1.   Build the larger number.
2.   Take away enough to leave ten.
3.   How much more must be taken?
4.   Record the number of units left.

$$\begin{array}{r} 14 \\ -\ 8 \\ \hline \end{array} \qquad \begin{array}{r} 14 \\ -\ 9 \\ \hline \end{array} \qquad \begin{array}{r} 14 \\ -\ 5 \\ \hline \end{array} \qquad \begin{array}{r} 14 \\ -\ 7 \\ \hline \end{array}$$

16

SUBTRACT (TAKE AWAY)

PREREQUISITE: Experience with the base ten strip past 200. Hungry Bug Subtraction. Played Go Broke.

MATERIAL: Base ten blocks.

PROCEDURE: Build the larger number. Take the smaller away (circled). Write the remainder.

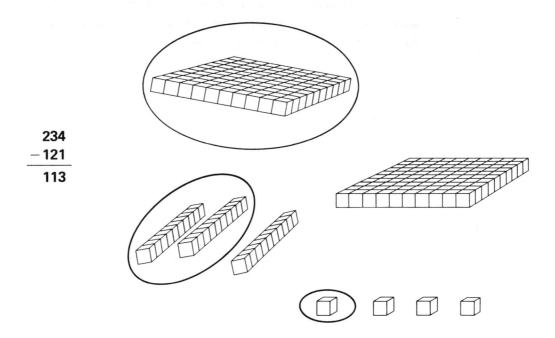

$$\begin{array}{r} 234 \\ -121 \\ \hline 113 \end{array}$$

Build these; remove the smaller number. Write the remainder

1. $\begin{array}{r} 146 \\ -\ 25 \\ \hline \end{array}$ 2. $\begin{array}{r} 146 \\ -\ 26 \\ \hline \end{array}$ 3. $\begin{array}{r} 146 \\ -\ 24 \\ \hline \end{array}$

Find some like this in your book. Build them. Record results.

SUBTRACT (COMPARISON)

PREREQUISITE: Experience with the base ten strip to 100. Addition with trading. Game of Go Broke.

MATERIAL: Base ten blocks.

PROCEDURE: (Subtraction asks a question with different interpretations. Two of these questions are [1] "take away" and [2] comparison.) Build the larger number. Place the blocks that represent the smaller number on top of the larger and record the part not covered.

   65
 −32
  33

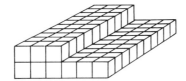
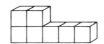

When possible cover units with units and longs with longs, etc. Sometimes it is not possible:

   65
 − 37
  28

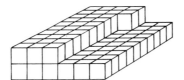
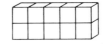

Build each of these: (Place the smaller number of blocks on the larger. Record what is not covered.)

$$52 \quad\quad 52 \quad\quad 52 \quad\quad 52$$
$$-21 \quad\quad -22 \quad\quad -23 \quad\quad -24$$

18

MORE SUBTRACT (COMPARISON)

PREREQUISITE: Experience with the base ten strip to at least 200. Addition with trading. Game of Go Broke.

MATERIAL: Base ten blocks.

PROCEDURE: Build the two numbers. Place the smaller number of blocks on top of the larger and record the part not covered.

$$\begin{array}{r} 145 \\ -\ 82 \\ \hline 63 \end{array}$$
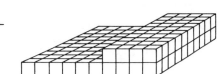 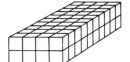

Build each of these. Place the smaller number of blocks on the larger. Record what is not covered.

1. $\begin{array}{r} 145 \\ -\ 83 \\ \hline \end{array}$

2. $\begin{array}{r} 145 \\ -\ 84 \\ \hline \end{array}$

3. $\begin{array}{r} 145 \\ -\ 85 \\ \hline \end{array}$

4. $\begin{array}{r} 145 \\ -\ 86 \\ \hline \end{array}$

Find more like this in your book. Build and record.

<u>SUBTRACTION (REGROUPING)</u>

PREREQUISITE: Experience with base ten strip past 100. Played Go Broke. Subtraction by comparison.

MATERIALS: Your computer and base ten blocks.

PROCEDURES: (Subtraction asks a question with different interpretations. Two of these questions are [1] "take away" and [2] "comparison.") This is a development of "take away" aimed at making understandable the "Borrowing" algorithm. Build the larger number on your computer. Examine the units in the smaller number; can you take away the units needed? (No) Look at Step 2.

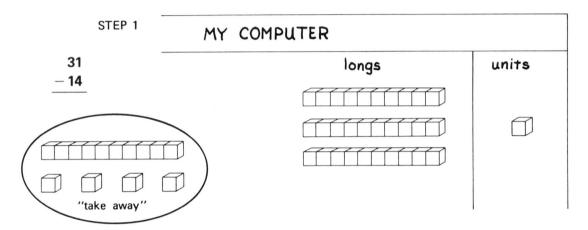

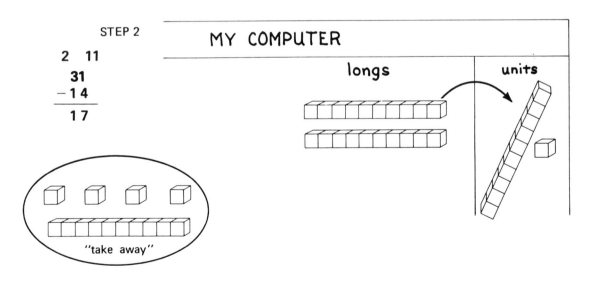

Find more problems in your book like these. Build, record.

1. 43
   −14
   ___

2. 43
   −15
   ___

3. 43
   −16
   ___

4. 43
   −17
   ___

5. 43
   −18
   ___

6. 43
   −19
   ___

SUBTRACTION ACROSS ZEROES

PREREQUISITE: Experience with the subtraction algorithm (or subtraction with regrouping).

MATERIAL: Base ten blocks and the computer mat.

PROCEDURE: (Subtraction asks a question with different interpretations. Two of these questions are [1] "take away" and [2] comparison.) This is further development of "take away" aimed at making understandable the borrowing algorithm. Build the larger number on your computer. Can you take seven units from one unit? (NO)

Exchange a flat for ten longs and one of the longs for ten units. (See step 2) Now "take away" can be done.

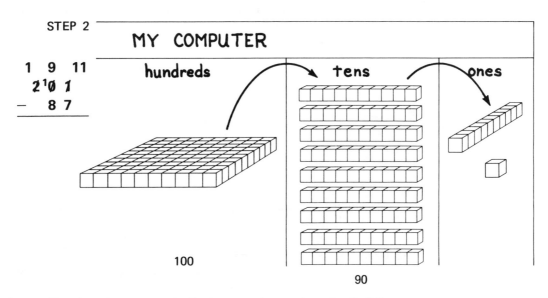

Find more like these in your book. Exchange and record results. Build!

	1. 301 − 46	2. 302 − 47	3. 303 − 48	4. 304 − 49	5. 305 − 50

RECTANGULAR ARRAYS (MULTIPLICATION)

PREREQUISITE: Experience with addition.

MATERIALS: Base ten blocks, square centimeter paper, crayons. (Square centimeter paper pattern inside front cover.)

PROCEDURE: Build 2 X 3. Push them together to make a rectangle. Color a region on the square centimeter paper on which this array would fit.

The multiply sentences for this array are: 2 X 3 = 6   and   3 X 2 = 6

Build 2 X 5. Push them together to make a rectangle. Color a region on the square centimeter paper on which this array would fit.

2 X 5 = 10

5 X 2 = 10

Cover these regions with units. Write the two multiply sentences for each.

The length and width are the FACTORS.

The number of squares is the PRODUCT.

Build these products. Write the two sentences about each.

1.   **4 X 5**   2.   **3 X 6**   3.   **2 X 6**   4.   **4 X 6**

22

MULTIPLES ON THE COUNTING STRIP

PREREQUISITE: Experience with the counting strip by ones in base ten past 100. Rectangular arrays.

MATERIALS: Base ten blocks, computer mats, adding machine tape for the strip, crayons,and a 100 chart for each multiple strip a child writes. (100 chart pattern on the inside back cover)

PROCEDURE: To make a multiple of 3 strip and chart, put 3 units on the computer and record 3 on the strip; put out another 3 units and record 6; put out another 3 units and record 9 on the strip; put out another 3 units, trade 10 units for a long and record 12 on the strip. Continue putting out 3 more until 100 is passed.

When the counting strip is complete, color in the corresponding number square on the 100 chart for each number on the strip. What is the pattern?

Each multiple will make a different pattern. Do a strip and chart for the following multiples: 2, 4, 5, 6, 7, 8, 9, 10 (Be sure to do 10)

Strip

3
6
9
12
15
18
21
.
.
.
.
.
.
.
.
99
102

MY COMPUTER

tens | ones

1	2	3	4	5	6	7	8	9	10
11	12	13	14	15	16	17	18	19	20
21	22	23	24	25	26	27	28	29	30
31	32	33	34	35	36	37	38	39	40
41	42	43	44	45	46	47	48	49	50
51	52	53	54	55	56	57	58	59	60
61	62	63	64	65	66	67	68	69	70
71	72	73	74	75	76	77	78	79	80
81	82	83	84	85	86	87	88	89	90
91	92	93	94	95	96	97	98	99	100

MULTIPLICATION TABLE

PREREQUISITE: Making the multiple strips and rectangular arrays.

MATERIALS: A blank multiplication table, white units, 10 longs, square centimeter paper, scissors.

PROCEDURE:
1. Build the products and push together to make an array.
2. Cut an array of square centimeter paper on which the white units fit.
3. Remove the blocks and trade when possible for longs.
4. Record the result on the paper array.
5. Record result on the multiplication table.

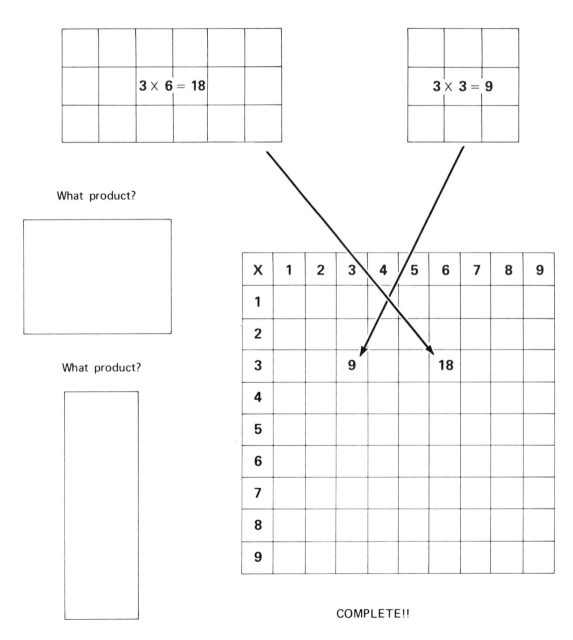

3 × 6 = 18

3 × 3 = 9

What product?

What product?

X	1	2	3	4	5	6	7	8	9
1									
2									
3			9			18			
4									
5									
6									
7									
8									
9									

COMPLETE!!

DIVIDE ASKS TWO QUESTIONS—DO IT BOTH WAYS

PREREQUISITE: Counting strips with multiples.

MATERIALS: Base ten blocks.

PROCEDURE: Take the problems of division to ask about 6.

"DIVVY UP"   *"MEASURED OUT"*

6 divided by 1 = 6

"Divvy Up" asks: "If we arrange six into one pile, how many in the pile?

"Measure Out" asks: "If we measure out one at a time, how many piles?" (How many ones in six?)

6 divided by 2 = 3

"Divvy Up" asks: "If we separate six into two piles, how many in each pile?

"Measure Out" asks: "If we measure out two at a time, how many piles?" (How many 2's in 6?)

6 divided by 3 = 2

"Divvy Up" asks: "If we separate six into three piles, how many in each pile?"

"Measure Out" asks: "If we measure out three at a time, how many piles?" (How many 3's in 6?)

6 divided by 4 = 1   R = 2

"Divvy Up" asks: "If we separate six into four piles, how many in each pile?"

"Measure Out" asks: "If we measure out four at a time, how many piles?" (How many 4's in 6?)

6 divided by 5 = 1   R = 1

"Divvy Up" asks: "If we separate six into five piles, How many in each pile?"

"Measure Out" asks: "If we measure out five at a time, how many piles?" (How many 5's in 6?)

6 divided by 6 = 1

"Divvy Up" asks: "If we separate six into six piles, how many in each pile?"

"Measure Out" asks: "If we measure out six at a time, how many piles?" (How many 6's in 6?)

25

MORE RECTANGULAR ARRAYS

PREREQUISITE: Experience with addition.

MATERIALS: Base ten blocks, mostly units and longs, square centimeter paper, crayons.

PROCEDURE: Show the required number with units. Build as many rectangles as possible. Record the multiplication and division sentences for each.

Fact family for 9

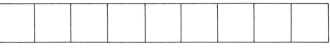

1 X 9 = 9   **9 X 1 = 9**
9 ÷ 1 = 9   **9 ÷ 9 = 1**

3 X 3 = 9
9 ÷ 3 = 3

Cover with units. What fact family? _____

_____ X _____ = _____   _____ X _____ = _____

_____ ÷ _____ = _____   _____ ÷ _____ = _____

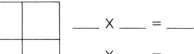

_____ X _____ = _____

_____ X _____ = _____

_____ ÷ _____ = _____

_____ ÷ _____ = _____

Cover with units. What fact family? _____

_____ X _____ = _____   _____ X _____ = _____

_____ ÷ _____ = _____   _____ ÷ _____ = _____

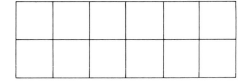

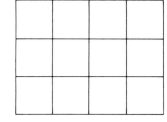

_____ X _____ = _____   _____ X _____ = _____

_____ ÷ _____ = _____   _____ ÷ _____ = _____

_____ X _____ = _____

_____ X _____ = _____

_____ ÷ _____ = _____

_____ ÷ _____ = _____

PRODUCTS WITH ONE DIGIT

PREREQUISITE: Rectangular arrays, multiplication table.

MATERIALS: Your computer, base ten blocks, paper and pencil.

PROCEDURE: Build the problem: **24**   by building three piles of 24:
                                   X 3

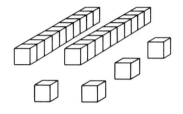 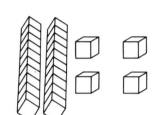 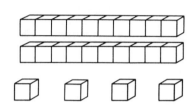

1. Pick up units.
2. Trade if you can.
3. Put on the computer.
4. Record.
5. Pick up longs.
6. Trade if you can.
7. Put on your computer.
8. Record.
9. Add to get the final product.

$$\begin{array}{r} 24 \\ \times\ 3 \\ \hline 12 \\ 60 \\ \hline 72 \end{array}$$

MY COMPUTER

hundreds	tens	ones

Do these examples the same way.

$$\begin{array}{cr} 1. & 43 \\ & \times\ 2 \\ \hline \end{array} \qquad \begin{array}{cr} 2. & 43 \\ & \times\ 3 \\ \hline \end{array} \qquad \begin{array}{cr} 3. & 43 \\ & \times\ 4 \\ \hline \end{array} \qquad \begin{array}{cr} 4. & 43 \\ & \times\ 5 \\ \hline \end{array}$$

Find more examples like these in your book.

PRIMES

PREREQUISITE: Rectangular arrays.

MATERIALS: Base ten blocks.

PROCEDURE: Prime numbers make ONLY long, unit-wide rectangles (exactly two factors). Numbers that can be made into more than one arrangement are not prime.

One is not prime; only one factor.   **1 X 1**

TWO is the only even prime.   **1 X 2 (or 2 X 1)**

THREE IS PRIME: **1 X 3 (or 3 X 1)**

Four is not prime: **1 X 4 (or 4 X 1)** and **2 X 2**

FIVE is prime: **1 X 5 (or 5 X 1)**

Build arrays for the numbers through 25, List the primes:

$$\{2, 3, 5, \text{\_\_\_}, \text{\_\_\_}, \text{\_\_\_}, \ldots\}$$

The other numbers (except 1) are called COMPOSITES.

List the composites between 1 and 26

$$\{4, \text{\_\_\_}, \text{\_\_\_}, \ldots\}$$

PRODUCTS WITH A ONE-DIGIT FACTOR

PREREQUISITE: Experience with rectangular arrays.

MATERIALS: Base ten blocks, computer, paper, pencil.

PROCEDURE: Build four piles of 436.

436
× 4

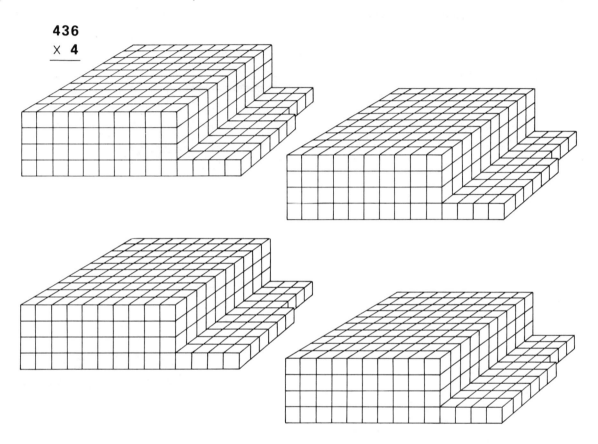

1. Collect the units and trade,
   and record.   ⟶   24
2. Collect the longs and trade,
   and record.   ⟶   120
3. Collect the flats and trade,
   and record.   ⟶ 1600
4. Add them all together to find the product. ⟶ 1744

Find more problems like this in your book. For each problem: Build, collect units first, and trade; then longs and trade; and finally flats and trade. Add them all together to get the product.

1.	345	2. 2.	345	3.	345	4.	345	5.	345
	× 3		× 4		× 5		× 2		× 6

QUOTIENTS WITH ONE-DIGIT DIVISORS

PREREQUISITE: Products with a one-digit factor.

MATERIALS: Base ten blocks, as many pads (or sheets of paper) as the number which is the divisor.

PROCEDURE: What is 639 divided by three?

1. Build **639**

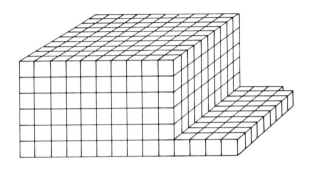

2. Since we are dividing by three, arrange 3 mats.
   - a. How many flats on each? (2)   Any left? NO.
   - b. How many longs? (1)   Any left? NO.
   - c. How many units? (3)   Any left? NO.

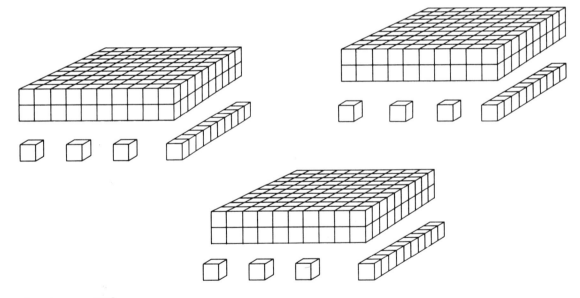

Quotient: **213**

Find more in your books (these have remainders)

Try   **639 ÷ 2**   also   **639 ÷ 4**

30

DIVISION HAS MORE THAN ONE INTERPRETATION

PREREQUISITE: Quotients with one-digit divisors.

MATERIAL: Base ten blocks

PROCEDURE: Children will usually think of "Divvy Up" when asked a division question. They need to be lead through the process of "Measure Out" with small dividends so the eventual long division algorithm will make sense.

PROBLEM: Use "Measure Out" division to solve: **120 divided by 30**

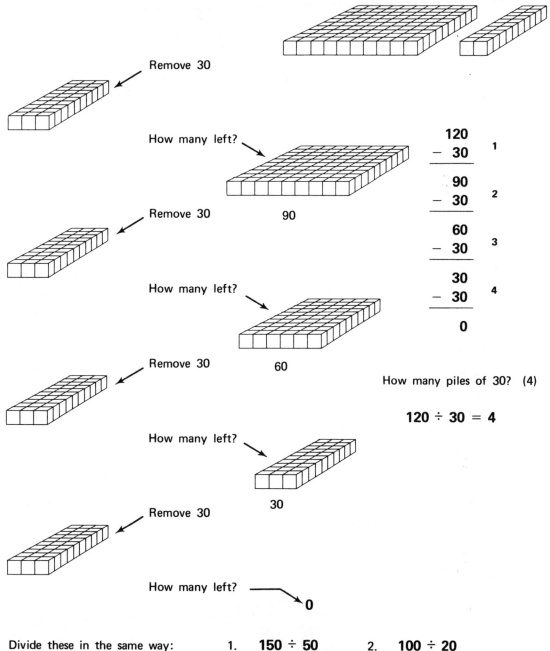

Remove 30

How many left?

Remove 30   90

How many left?

Remove 30   60

How many left?

Remove 30   30

How many left? ⟶ 0

$$\begin{array}{r} 120 \\ -\ 30 \\ \hline 90 \\ -\ 30 \\ \hline 60 \\ -\ 30 \\ \hline 30 \\ -\ 30 \\ \hline 0 \end{array}$$

1

2

3

4

How many piles of 30? (4)

120 ÷ 30 = 4

Divide these in the same way: 1. **150 ÷ 50**   2. **100 ÷ 20**

MORE QUOTIENTS WITH ONE-DIGIT DIVISORS

PREREQUISITE: Quotients with one digit divisors (without leftovers) with blocks.

MATERIALS: Base ten blocks, pencil, and paper. As many mats as the number in the divisor.

436 divided by 3   Write: $3\overline{)436}$ Build **436**
(Divvy Up)

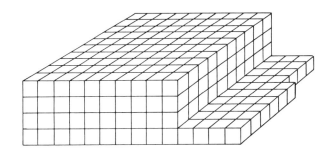

Questions:

1. Into how many piles? (3)

2. With 4 flats, how many to each pile?

3. How many flats used?

4. How many flats left?

5. One flat is 10 longs plus the 3 longs already there.

6. 13 longs into 3 piles, how many longs in each pile? (4)

7. How many longs used?

8. How many longs left?

9. One long is 10 units plus the 6 units already there.

10. 16 units into 3 piles, how many units? (5)

11. How many used?

12. How many left?

13. Write the numeral for the blocks in each pile.

```
        1 4 5
    3 )4 3 6
     - 3
      1 3
     -1 2
        1 6
       -1 5
          1
```

(in each pile)

Check:

```
    1 4 5
  X   3
     1 5
   1 2 0
   3 0 0
   4 3 5
  +   1
   4 3 6
```

Remainder:

145

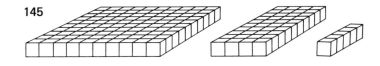

32

PRODUCTS WITH TENS

PREREQUISITE: Rectangular arrays with units.

MATERIAL: Flats from the base ten blocks.

PROCEDURE: Children need to be told that they are going to discover a secret of the zeroes; watch for patterns. Have children build each product.
NOT FOR CHILDREN—(do the real building with them) but to communicate to the teacher the flat is represented by a square:

will be

Have them find one piece that is 10 X 10

10 X 10 = 100

Build 10 X 20

10 X 20 = 200

Build 10 X 30

10 X 30 = 300

Build 20 X 20

20 X 20 = 400

What is the secret?   Can you use it to find: **20 X 50** = _____   Build to check.

20 X 30 = _____   **30 X 30** = _____   **20 X 30** = _____

NAME THE PARTIAL PRODUCTS

PREREQUISITE: Rectangular arrays and the Secret of the Zeroes.

MATERIALS: Base ten blocks.

PROCEDURE: Cover this region with as many flats as possible, then as many longs, and finally fill in the corner with units. Place your flat in one corner.

What is the problem?   _____ X _____ = __?__

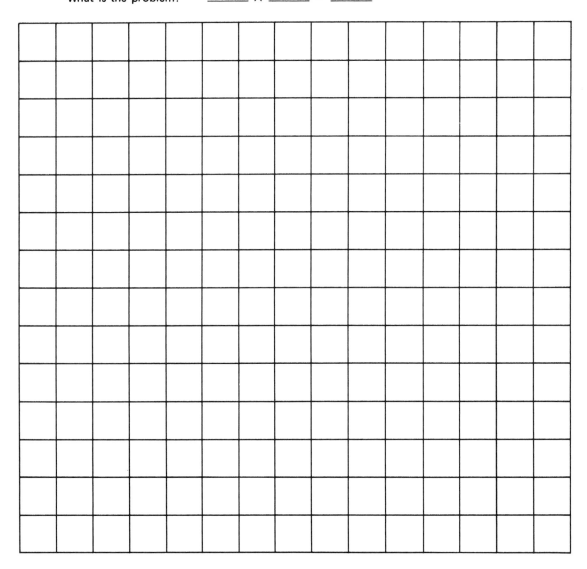

Remove your flat and label its region   **10 X 10 = 100**

Find and label these three more regions: **10 X 4,   10 X 5,   4 X 5**

Sum the four regions to find the product: _____ + _____ + _____ + _____ = _____

34

MAKE IT GROW

PREREQUISITE: Rectangular arrays and products with tens.

MATERIALS: Base ten blocks.

NOTE: Symbol for a long: and flat:

are [   ]
         long

(for teachers only; children build!!)

flat

Build: 10 X 10

Make it grow to 11 X 11   = 121
a flat, 2 longs, 1 unit

Make it grow to 11 X 12   = 132

Make it grow to 12 X 12   = 144

Make it grow to 12 X 13   = 156

13 X 13 = _____   (Guess. Then build and check.)

RECTANGULAR ARRAYS BUILD PRODUCTS

PREREQUISITE: Rectangular arrays with units and Make It Grow.

MATERIALS: Base ten blocks.

PROCEDURE: The arrays are to be built with the blocks.

The power of conceptualizing multiplication as an array cannot be over emphasized. A young child thinks of three times four as:

This is naturally extended to problems like 22 X 13 which is pictured in the accompanying photograph.

At first the child counts up the two flats, eight longs and six units and gives 286 as the product. However from this rectangular array, the algorithm can be understood by naming the four partial products that correspond to the four separate regions. Since they already know the product is the sum of units in the four regions, it is not necessary to formally name the distributive law. The algorithm becomes a scheme or pattern for making sure the four regions are computed and added. The scheme the author has used with children is a pattern of arrows:

$$\uparrow \quad \diagup\!\!\!\diagdown \quad \uparrow$$

When applied to 22 X 13:

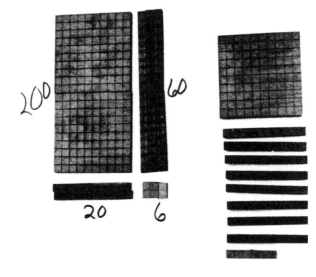

```
     22
   X 13

     6 = 3 X 2 and is ↑
    60 = 3 X 20 and is ↖
    20 = 10 X 2 and is ↗
   200 = 10 X 20 and is ↑

   286
```

After a child has learned to build the arrays and find the product by trading and recording, the four partial products can be formalized by asking:

Show me 3 X 2.   Record.

Show me 3 X 20.   Record.

Show me 10 X 2.   Record.

Show me 10 X 20.   Record.

I'll add up the numbers and you put the blocks together and see if we agree. This should be done until a child exclaims, "I can do it without the blocks!" (Caution: Propose two digit numbers with small numbers in the units place so the supply of blocks will be adequate. Children will generalize from the small ones and extend the products to larger two place numbers.)

36

MORE RECTANGULAR ARRAYS BUILD PRODUCTS

PREREQUISITE: Rectangular arrays.

MATERIALS: Base ten blocks.

PROCEDURE: All of the suggested problems are to be built with blocks. The diagrams below are to help teachers see the sequencing of activities and questions needed to develop understanding of the multiplication algorithm.

Again the flat is represented by

the long by

and unit

Build 23 X 32 and separate it into four regions to represent the four partial products.

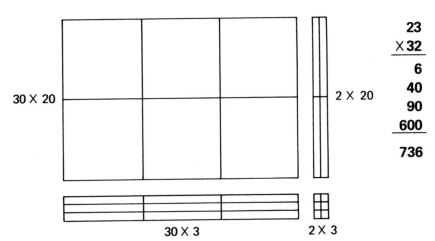

$$30 \times 20 \qquad 2 \times 20$$

$$30 \times 3 \qquad 2 \times 3$$

```
    23
  X 32
     6
    40
    90
   600
   736
```

The problems below can be built with the blocks to further practice the strategy above:

43	52	13	61	42	12
X 21	X 32	X 42	X 22	X 34	X 72

RECTANGULAR ARRAYS INTERPRET DIVISION, TOO

PREREQUISITE: Rectangular arrays.

MATERIALS: Base ten blocks.

PROCEDURE: The purpose of this activity is to show the relationship between multiplication and division that can be understood from the rectangular array.

1.   What is 276 divided by 23?   Take out 276. Build an array 23 wide.

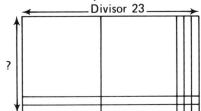

What is the other dimension?

It is the quotient.

276 ÷ 23 = 12

The above diagram (and block array built by child) represents both multiplication and division:

PRODUCT	=	FACTOR	X	FACTOR
276		**23**		**12**
DIVIDEND	÷	DIVISOR	=	QUOTIENT

2.   Sometimes there is a remainder. What is 272 ÷ 11?

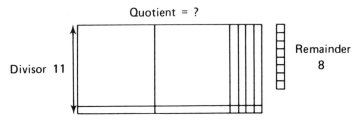

Either dimension can be the divisor, the other is the quotient. What is left is the remainder.

272 ÷ 11 = 24   R = 8

Solve the following division problems by building arrays with one dimension the divisor.

    1.   **738 ÷ 23**   3.   **448 ÷ 32**

    2.   **149 ÷ 12**   4.   **448 ÷ 31**

DIVISION USING MULTIPLES OF 10 AND 100

PREREQUISITE: Division by successive subtraction.

MATERIALS: Base ten blocks.

PROCEDURE for PROBLEM: Measured division.

13)‾1476   What is 10 X 13?   130

What is 100 X 13?

Build:

1300

Remove 100 13's

```
        100
  13 )1476
     -1300
       176
```

```
     10
    100
13 )1476
   -1300
     176
     130
      46
```
Remove 10 13's

```
         3
        10
       100
  13 )1476
     -1300
       176
     - 130
        46
      -  39
         7
```

1 X 13 = 13
2 X 13 = 26
3 X 13 = 39
4 X 13 = 52

Remove 3 13's

Quotient 113   Remainder 7

DECIMAL SPIN ONE TENTHS AND HUNDREDTHS

PREREQUISITE: Experience using base ten blocks with whole numbers.

ASSUMPTION: Let the flat be ONE:

MATERIALS: 10 X 10 cm grid, two
dice marked:

0.1, 0.2, 0.3, 0.4, 0.5, 0.6

and

0.09, 0.08, 0.07, 0.06, 0.05, 0.04

and

Base ten blocks.

RULES: Each player throws the dice and the one with the largest sum goes first. Play proceeds in a clockwise direction.

For example, player 1 throws 0.09 and 0.1; he places on his mat:

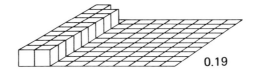

0.19

On player 1's next throw, he throws 0.3 and 0.08 and gets out:

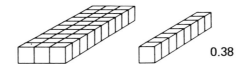

0.38

ADDING IT TO HIS MAT, TRADING WHEN NECESSARY, HE HAS:

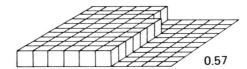

0.57

Play continues until one player has EXACTLY ONE. Only one die can be used after 0.09 is reached.

40

DECIMALS: SPIN ONE—TENTHS, HUNDREDTHS, THOUSANDS

PREREQUISITE: Play with the Spin One Game for hundredths.

ASSUMPTION: Let the block be ONE:
Then, the flat is 0.1
Long is 0.01
Unit is 0.001

$1 =$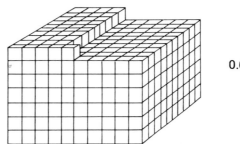

MATERIALS: A 10 × 10 grid,
Three dice marked:
0.1, 0.2, 0.3, 0.4, 0.5, 0.6
0.09, 0.08, 0.07, 0.06, 0.05, 0.04
0.001, 0.002, 0.003, 0.004, 0.005, 0.006
Base ten blocks.

RULES: Play proceeds as in the previous game except the ONE is a block. For example, if player 1 spun 0.6 and 0.04 and 0.001, his board would look like this:

0.641

On the next spin for player 1's turn, if he spins 0.1 and 0.04 and 0.002

$$
\begin{array}{r}
0.641 \\
+ \ 0.142 \\
\hline
0.783 \quad \text{etc.}
\end{array}
$$

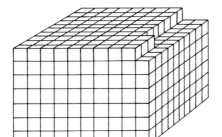

LOSE ONE

Subtracting Decimals

PREREQUISITE: SPIN ONE games.

GAME: The same ideas as in the GO BROKE game for whole numbers except that either a flat or a block is agreed upon for the whole, which each player begins the game with. To have players record as they go and subtract to verify their block play reinforces the idea and causes students to conclude: "The decimal points must be in line!"

MATERIALS THE SAME AS FOR THE PREVIOUS TWO GAMES.

The recording would look like this:

If 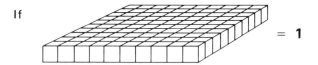 = **1**

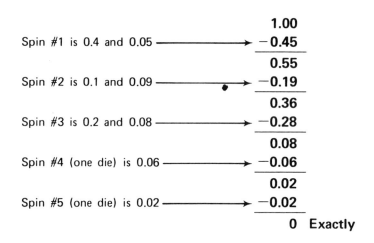

Spin #1 is 0.4 and 0.05 ⟶

Spin #2 is 0.1 and 0.09 ⟶

Spin #3 is 0.2 and 0.08 ⟶

Spin #4 (one die) is 0.06 ⟶

Spin #5 (one die) is 0.02 ⟶

$$
\begin{array}{r}
1.00 \\
-0.45 \\
\hline
0.55 \\
-0.19 \\
\hline
0.36 \\
-0.28 \\
\hline
0.08 \\
-0.06 \\
\hline
0.02 \\
-0.02 \\
\hline
0 \ \ \text{Exactly}
\end{array}
$$

WIN

42

COMPARING DECIMALS

PREREQUISITE: SPIN a whole and LOSE ONE.

MATERIALS: Base ten blocks and whole number comparison pictures pages

ASSUMPTION: Let a block be ONE
                  A flat is 0.1
                  A long is 0.01
                  A unit is 0.001

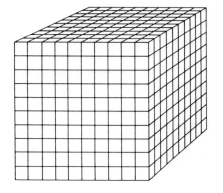

With these assumptions use the pictures on pp. 8-11. Build each number and write the decimal name. Rank each from smallest to largest.

THE DECIMAL POINT MAKES THE DIFFERENCE!

0.001 < 0.010 < 0.100 < 1.000

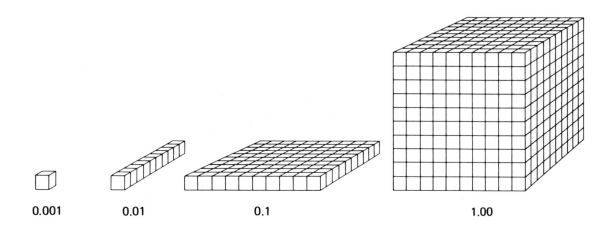

0.001   0.01   0.1   1.00

43

ADDITION OF DECIMALS

PREREQUISITE: Spin a ONE and Lose One games.

MATERIAL: Base ten blocks.

ASSUMPTION: Let a block be ONE

PROBLEM: **1.394**
           + .635

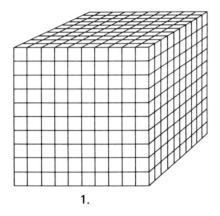

1.

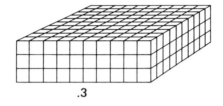

.3

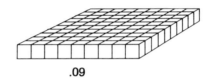

.09

.004

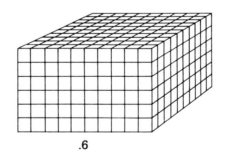

.6

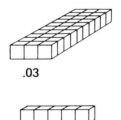

.03

.005

Build, add, trade, record. Find some in your book. Check with blocks.

1.   **2.076**
     + .847

2.   **1.437**
     + .563

44

DECIMAL PRODUCTS

PREREQUISITE: Rectangular arrays with whole numbers.

MATERIAL: Base ten blocks.

ASSUMPTION: The length of a long is ONE for the factors.

Since the long is ten tenths, a flat is the product of 10 tenths by 10 tenths or 100 hundredths or ONE.

PROBLEM: **2.3 X 1.4**

BUILD:

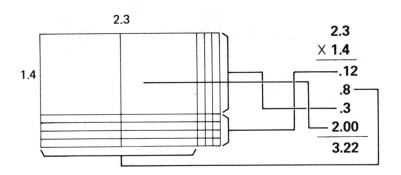

PROBLEM: **4.2 X 2.3**

BUILD:

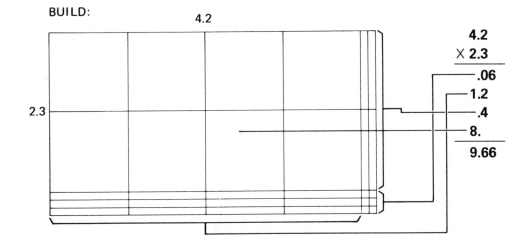

Build and solve these decimal products:

1. **2.3 X 1.4**   3. **2.5 X 4.2**

2. **4.1 X 3.2**   4. **0.5 X 4.2**

45

DECIMAL DIVISION

PREREQUISITE: Decimal products.

MATERIAL: Base ten blocks.

ASSUMPTION: A flat is ONE in the dividend a long is 0.1

PROBLEM: **1.56** divided by **1.3**

BUILD: 1.56

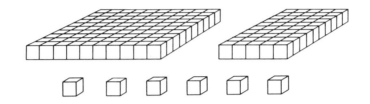

Rearrange

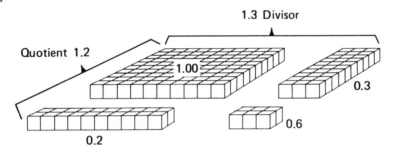

$$1.56 \div 1.3 = 1.2$$

Build to find these quotients, build arrays:

1.   **1.56 ÷ 2.6**   3.   **1.56 ÷ 5.2**

2.   **1.56 ÷ 3.9**   4.   **1.56 ÷ 7.8**

Find more decimal division in your book. Build and solve.

46

© ACTIVITY RESOURCES COMPANY, INC., Box 4875, Hayward, CA 94545

METRIC LENGTH

PREREQUISITE: Addition activities.

MATERIALS: Base ten blocks, metric ruler (may be cut out from below).

PROCEDURE: Measure the edge of a long; it is 1 centimeter.
Measure the length of a long; it is 1 decimeter.

Cover this shape with base ten blocks. Use the centimeter ruler below to measure the distance around (Perimeter is the sum of the lengths of the line segments that make a region.) Also measure with the decimeter ruler and express the perimeter to the nearest tenths of a decimeter. What is the relationship?

Make some more shapes by drawing around regions that you create with the blocks. Save your shapes for the next page when area questions will be asked.

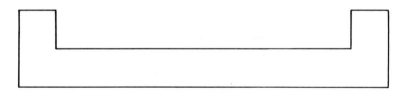

The blocks themselves can serve as measuring devices for perimeter if you think of the distance around.

CENTIMETER RULER

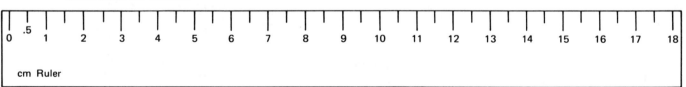

DECIMETER RULER

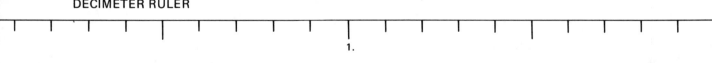

47

METRIC AREA

PREREQUISITE: Rectangular arrays.

MATERIALS: Base ten blocks, square centimeter paper, scissors.

PROCEDURE: Area is the number of squares needed to cover a region.

Look back at pages 37 and 38; the factors are the dimensions of rectangles.

The products are the number of squares needed to cover. Each square is a square centimeter (symbol cm^2).

Use the regions you drew for page 47. Cut regions to fit them from cm^2 paper. How many squares are needed to cover?

Can you make a rule?

Your blocks are scored in cm^2 so you can measure area by examining the number of squares on top. The units' faces are cm^2. The flats are square decimeters; the symbol is dm^2.

How many cm^2 are in 1 dm^2? _____

Cover these with base ten blocks. Express as cm^2 and dm^2 to the nearest hundredth.

_____ cm^2

_____ dm^2

METRIC VOLUME

PREREQUISITE: Addition and subtraction activities.

MATERIALS: Base ten blocks, cardboard, scissors, tape.

PRODEDURE: Each unit is one centimeter cubed (1 cm^3); each 1000 block is one decimeter cubed (1 dm^3). Use the base ten blocks to build these rectangular solids and others like them. How many cm^3 in each?

The number of cm^3 is the VOLUME.

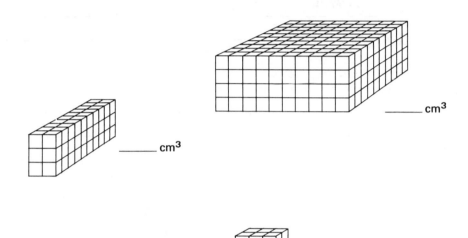

_____ cm^3

_____ cm^3

_____ cm^3

Each thousand block has a volume or capacity of 1 litre. Use five pieces of cardboard taped to fit the outside of a block; the box has a capacity of 1 litre. Since 1000 cm^3 are needed to fill the box and one litre is 1000 millilitres (ml), then the capacity of 1 cm^3 = 1 ml.

Find some small boxes and fill them with cm^3. This number will be the volume; it will also indicate the number of millilitres (ml).

Describe the Box	Number of cm^3	Number of ml
Match Box		

49

METRIC MASS

PREREQUISITE: Rectangular arrays, volume.

MATERIALS: Base ten blocks, objects to balance, simple pan balance.

PROCEDURE: Each unit is 1 gram. Choose some objects in the classroom. List them in this table. Balance with units. Record in grams.

OBJECT	MASS IN GRAMS
pencil	
crayon	
scissors	
eraser	
five paper clips	

unit = 1 gram

One cubic centimeter (1 cm^3) of water has a mass of 1 gram. The white units also have a mass of 1 gram.

If you filled the litre box with units, how many would it take. What is the mass? (1000 grams are a kilogram)

If the mass of your object is more than you can balance with the units you have, how can you solve your problem?

HINT: Find the mass of something else you have many of, like dominoes, then use the dominoes as mass pieces.

Many producers of cereal and crackers label their products with metric units; find some of these and check with the suggestions here using your base ten blocks.

50

METRIC UNITS ARE INTERRELATED

PREREQUISITE: Activities with volume, capacity, and mass.

MATERIALS: Base ten blocks.

PROCEDURES: One unit is the unit for

One unit is the unit for	VOLUME	LIQUID CAPACITY	MASS
1 unit	1 cm^3	1 ml	1 g
1000 units	1 dm^3	1 litre	1 kilogram

IF the figures below are built of white units, record the volume, capacity, and mass.

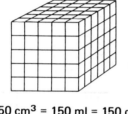

150 cm^3 = 150 ml = 150 g

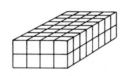

64 cm^3 = _____ ml = _____ g

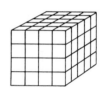

_____ cm^3 = _____ ml = _____ g

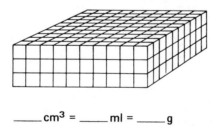

_____ cm^3 = _____ ml = _____ g

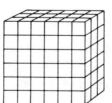

_____ cm^3 = _____ ml = _____ g

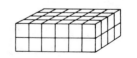

_____ cm^3 = _____ ml = _____ g

PREREQUISITE: Experience with fractions and decimals.

MATERIALS: Base ten blocks, square centimeter paper, scissors.

PROCEDURE: Build a small structure. Multiply each dimension by 2 and build.

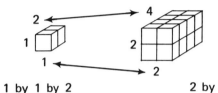

1 by 1 by 2   2 by 2 by 4

EDGE RATIO: $\dfrac{1}{2}$ or $\dfrac{2}{4}$

Make jackets out of square centimeter paper: (Children need to experience cutting a jacket to exactly fit their structure to understand that area is the number of squares to cover.)

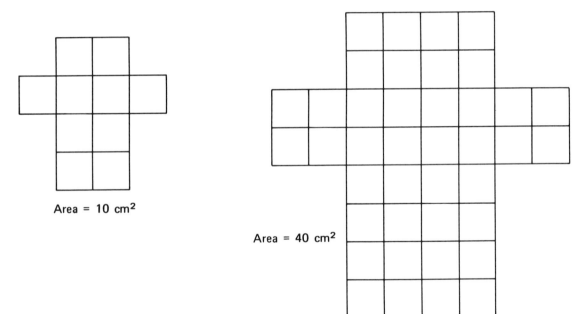

Area = 10 cm^2

Area = 40 cm^2

AREA RATIO = $\dfrac{10}{40}$ = $\dfrac{1}{4}$ = $\left(\dfrac{1}{2}\right)^2$

Fill each jacket with white cubes. The number used is the VOLUME.

VOLUME RATIO = $\dfrac{2}{16}$ = $\dfrac{1}{8}$ = $\left(\dfrac{1}{2}\right)^3$

Build another small structure. Double the dimensions. Predict areas and volumes. Check by making jackets. What are the ratios?

52

PREREQUISITE: Rectangular arrays as products.

MATERIALS: Base ten blocks.

Build these into squares:

4 units

$\sqrt{4} = 2$

9 units

$\sqrt{9} = 3$

16 units

$\sqrt{16} = 4$

20 units

Since
$$16 < 20 < 25$$
$$4 < \sqrt{20} < 5$$

*Therefore
$$\sqrt{20} \approx 4\frac{4}{9} \approx 4.4$$

(see page 54 for more)

25 units

$\sqrt{25} = 5$

What are the perfect squares between 25 and 121? Build them and write their square root.

Approximate $\sqrt{32}$ $\sqrt{55}$ $\sqrt{41}$ to the nearest tenth.

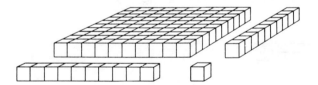

121

$\sqrt{121} = 11$

* "$\approx$" means approximate.

APPROXIMATE SQUARE ROOTS

PREREQUISITE: Square roots.

MATERIALS: Base ten blocks.

NOTE: Flats are   longs are   units ☐

Suppose there is a region with an area of 496 square units; how long is each edge? (What is the square root of 496?)

Build the largest possible square:

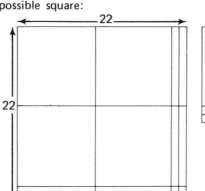

$$\sqrt{496} \approx 22.3$$

$$\left(\frac{12}{45} \approx .3 \right)$$

45 more units are needed to make (23)²

$$\frac{12}{45} \approx .3$$

Build and approximate the square roots of the following:

1.   **162**   2.   **450**   3.   **500**

(Trade a flat for 10 longs)

54

PREREQUISITE: Experience with whole numbers using the base ten blocks; operations with signed numbers.

MATERIALS: Base ten blocks or Multibase blocks.

ASSUMPTION: A long is x   a flat is x^2

units are 1's

Base ten blocks make excellent models for algebraic representations. It is essential that the student understand that the ten relationship so far as a trade is concerned is no longer possible since "x" is variable. For that reason it is better to have multibase blocks for the algebra. Then, using a long in the different bases to be "x", it is obvious that no one can trade unless everyone can. Below is an illustration of the sum of two polynomials.

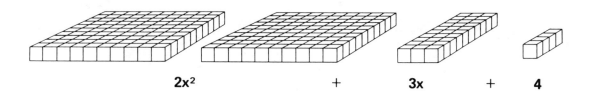

$2x^2$ + $3x$ +   4

$2x^2 + 3x + 4$
$\underline{x^2 + 2x + 3}$
$3x^2 + 5x + 7$

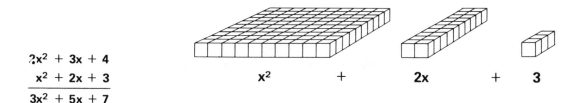

x^2 + $2x$ +   3

$$3x^2 + 5x + 7$$

Build and add:

1. $4x^2 + 7x + 2$
     $\underline{3x^2 + 1x + 4}$

2. $2x^2 + x + 77$
     $\underline{5x^2 + 3x + 1}$

55

© ACTIVITY RESOURCES COMPANY, INC., Box 4875, Hayward, CA 94545

PREREQUISITE: Operations with whole numbers using base ten blocks. Operations on signed numbers.

MATERIALS: Base ten or Multibase blocks.

ASSUMPTION: A long is x   then a flat is x^2

$\square$ = 1

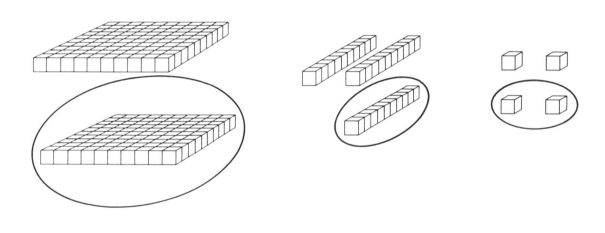

$$2x^2 \qquad + \qquad 3x \qquad + \qquad 4$$

Subtract $\qquad (\ x^2 \qquad + \qquad x \qquad + \qquad 2)$

$$\overline{\qquad x^2 \qquad + \qquad 2x \qquad + \qquad 2 \qquad}$$

Build, subtract, record:

1. $4x^2 + 6x + 7$
      $-(2x^2 + 3x + 2)$
      $\overline{\qquad\qquad\qquad\qquad}$

2. $5x^2 + 2x + 3$
      $-(2x^2 + \ x + 1)$
      $\overline{\qquad\qquad\qquad\qquad}$

ALGEBRA—PRODUCTS OF BINOMIALS

PREREQUISITE: Rectangular arrays for two two-place whole numbers.

MATERIALS: Base ten blocks or Multibase blocks.

ASSUMPTION: The length of a long is x. The factors describe the length and width; the area represents the product so the flat becomes x^2 in the product, the longs are x and the units ones.

Build and write the expanded product.

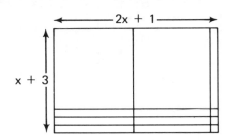

$(2x + 1)(x + 3)$

$2x^2 + 7x + 3$

Build and write the expanded product.

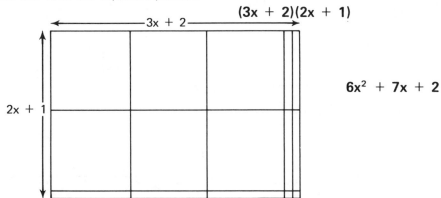

$(3x + 2)(2x + 1)$

$6x^2 + 7x + 2$

Build and write the expanded form:

1. $(2x + 3)(x + 5)$ 2. $(x + 5)(2x + 1)$

FACTORING TRINOMIALS

PREREQUISITE: Rectangular arrays.

MATERIALS: Base ten or Multibase blocks.

PROCEDURE: List the dimensions and the product for each of the following rectangular arrays.

a
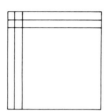

b
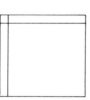

c
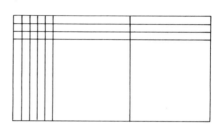

$$(x + 2)(x + 2) = x^2 + 4x + 4$$

d
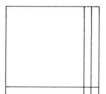

e

f
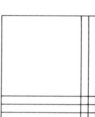

g
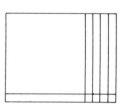

h
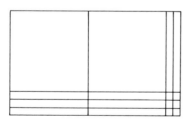

i
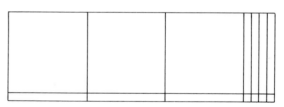

j
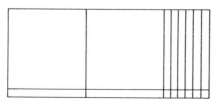

k
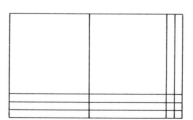

58

ALGEBRA—THE NEXT CUBE

PREREQUISITE: Rectangular arrays in algebra.

MATERIALS: Base ten or Multibase blocks.

ASSUMPTION: For linear dimensions the length of a long is x.

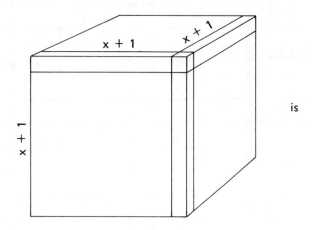

is

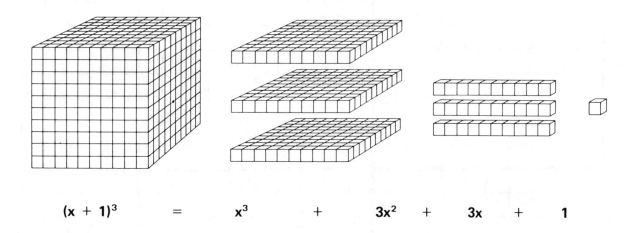

$$(x + 1)^3 \quad = \quad x^3 \quad + \quad 3x^2 \quad + \quad 3x \quad + \quad 1$$

Build: $(x + 2)^3$

BASE TEN PAPER

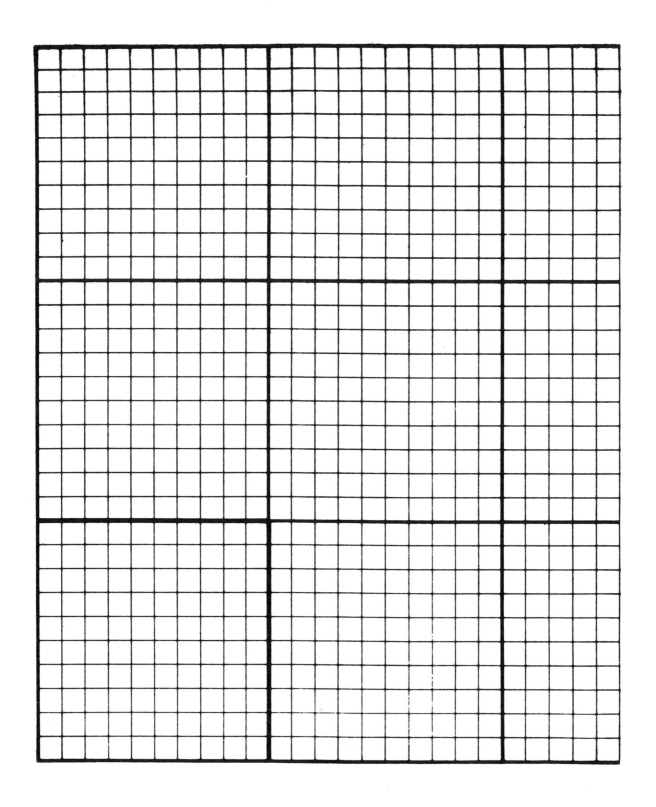